Hypothesis Testing by Example
Hands on approach using R

First Edition

Faye Anderson, MS, PhD

Contents

Preface   4

Chapter 1: Interpretation   5

Example 1: Steps of hypothesis testing 7

Chapter 2: One-sided vs Two-sided Hypothesis Testing
        8

Example 2: One Sample Test (one-sided)   8

Example 3: Two Groups (two-sided)   9

Example 4: Two Paired Groups (one-sided)   12

Example 5: Three Groups – No Order   13

Example 6: Three Ordered Groups   16

Chapter 3: Univariate and Multivariate Hypothesis
Testing   19

Example 7: Univariate Normality Test   19

Example 8: Multivariate Normality Test   21

Example 9: Henze-Zirklers Multivariate Normality Test
        22

Chapter 4: Parametric and Non-parametric Hypothesis
Testing   25

Example 10: ANOVA Test 26

Example 11: Kruskal-Wallis Non-parametric Test   29

Example 12: MANOVA Test   31

Veritas

Preface

Every statistical analysis needs hypothesis testing. The number of tests has been increasing. This book presents an overview of the concept of hypothesis testing and explains the different types and applications through examples. We start with the interpretation then proceed with the different flavors and when can they be used. The best way to benefit from my book is to do the examples and read the interpretation. The following table lists the twelve topics covered by the examples:

Example	Topic
1	Steps of hypothesis testing
2	One-sample test (one-sided/one-tailed)
3	Two Groups (two-sided)
4	Two Paired groups (one-sided)
5	Three Groups – No Order
6	Three Ordered Groups
7	Univariate Normality Test
8	Multivariate Normality Test
9	Henze-Zirklers Multivariate Normality Test
10	ANOVA Test
11	Kruskal-Wallis Non-parametric Test
12	MANOVA Test

As with my other books, R was selected because of its free accessibility. If you have any questions please feel free to post them in my Amazon author's blog.

Enjoy!

Chapter 1: Interpretation

Hypothesis testing assesses two mutually exclusive statements about the sampled data. The two mutually exclusive statements are called the null hypothesis (H_0) and the alternative hypothesis (H_1).

It might be odd to start the book with the interpretation, but this makes sense knowing that interpreting a hypothesis test result is not the tricky part. Once you have decided which test to conduct and what it really represents, its interpretation is straightforward just as in the table below.

p-value vs significance level (α)	Interpretation
p-value <= significance level e.g.; p-value = 0.01 & significance level = 0.05	The null hypothesis (H_0) is rejected. There is strong evidence against H_0.
p-value > significance level e.g.; p-value = 0.02 & significance level = 0.01	The null hypothesis (H_0) cannot be rejected. There is weak evidence against H_0.
p-value is very close to the significance level (also known as marginal p-value)	Conclude either way!

P-value $\leq \alpha \Rightarrow$ Reject H_0 at level α

P-value $> \alpha \Rightarrow$ Do not reject H_0 at level α

Once you conduct a hypothesis test, you get one important value: The p-value which is the probability that the data was sampled under the condition that the null hypothesis (H_0) is true. Therefore, the p-value has connection to sample size and sampling technique. This is a very important aspect but it is beyond the scope of this book.

Example 1: Steps of hypothesis testing

In statistics, in order to prove a hypothesis we assume the opposite and try to prove ourselves wrong!

A researcher measured blood pressure for hypertension patients who took his prescribed medicine for three months. He wants to test the hypothesis that the medicine has lowered the sample's average systolic blood pressure (bp) to 120. Here are the steps that needs to be followed:

1. Formulate the null hypothesis H_0: bp = 120, and the alternative hypothesis H_1: bp > 120

2. Set the significance level α (default value = 0.05).

3. Sample n values from the population that has N values. Assume that n = 30 patients.

4. Calculate p-value (Chapter 2 and Chapter 3).

5. Interpret the test based on the p-value compared to the significance level (Chapter 1).

However, you can also fit a model without conducting hypothesis testing first. Then check the model's F-test result which is included in R output for the model. Either way (testing before modeling or checking model's output), the assumption of data's normality should be checked prior to modeling. More on this in this book and my first book: Statistics by Exmaple.

Chapter 2: One-sided vs Two-sided Hypothesis Testing

Calling the test one-sided or one-tailed versus two-sided or two-tailed depends on the alternative hypothesis. If H_1 tests > or < then it's one-sided. If H_1 tests <> or $\neq$, it is two-sided.

Example 2: One Sample Test (one-sided)

Assume that we measured the lengths of ten plants (x). Is there evidence that the plants are longer than 5m?

H0: = 5

H1: > 5

```
x = c(2, 1.5, 3, 1.9,1.4, 2.2, 3.1, 1.89, 2.1,
1.88)
> summary(x)
   Min. 1st Qu.  Median    Mean 3rd Qu.    Max.
  1.400   1.882   1.950   2.097   2.175   3.100
> t.test(x, alternative="greater", mu=5)
         One Sample t-test
data:   a
t = -16.4221, df = 9, p-value = 1
alternative hypothesis: true mean is greater than
5
95 percent confidence interval:
 1.772954       Inf
sample estimates:
mean of x    2.097
```

Interpretation: The p-value is 1, which is greater than the significance level of 0.05. The null hypothesis (H₀) cannot be rejected. There is weak evidence against H₀. Hence, the mean length of the plants is greater than 5m. Please read more about this function by running the command help(t.test).

Example 3: Two Groups (two-sided)

```
data(sleep)
> dim(sleep)
[1] 20   3
head(sleep)
> head(sleep)
   extra group ID
1   0.7     1   1
2  -1.6     1   2
3  -0.2     1   3
4  -1.2     1   4
5  -0.1     1   5
6   3.4     1   6
```

```
help(sleep)
Student's Sleep Data
Description
Data which show the effect of two soporific drugs
(increase in hours of sleep compared to control)
on 10 patients.
Usage
sleep
Format
A data frame with 20 observations on 3 variables.

[, 1] extra  numeric    increase in hours of sleep
[, 2] group  factor     drug given
[, 3] ID     factor     patient ID
Details
The group variable name may be misleading about
the data: They represent measurements on 10
persons, not in groups.

> t.test(extra ~ group, data = sleep)

        Welch Two Sample t-test

data:  extra by group
t = -1.8608, df = 17.776, p-value = 0.07939
alternative hypothesis: true difference in means
is not equal to 0
95 percent confidence interval:
 -3.3654832  0.2054832
sample estimates:
mean in group 1 mean in group 2
          0.75            2.33
```

Interpretation: The sleep dataset stores two-group sleep data for 30 students. Each group followed a different prescription. By default, t.test tests that mu or average is zero. Hence in this case it tested the null hypothesis that the difference between sleep averages between the two groups (mu1-mu2) is zero (H_0: mu1-mu2 = 0). P-value is greater than the default significance level. Therefore, we reject H_0.

Example 4: Two Paired Groups (one-sided)

```
> group1 = c(91, 87, 99, 77, 88, 91)
> group2 = c(101, 110, 103, 93, 99, 104)
> t.test(group1,group2,alternative="less",
var.equal=TRUE)

        Two Sample t-test

data:  group1 and group2
t = -3.4456, df = 10, p-value = 0.003136
alternative hypothesis: true difference in means
is less than 0
95 percent confidence interval:
      -Inf -6.082744
sample estimates:
mean of x mean of y
 88.83333 101.66667
```

Interpretation: Since the p-value is very low, we reject the null hypothesis that the two groups' means are equal. The `var.equal=TRUE` assumed that the variances of the two groups are equal. This is also called the assumption of homogeneity of variance. If the two groups means are equal (p-value > 0.05) and their variances are equal then we can safely combine the two groups together as one group since we do not have enough evidence that they come from two different distributions.

Example 5: Three Groups – No Order

```
data(PlantGrowth)
head(PlantGrowth)
> head(PlantGrowth)
  weight group
1   4.17  ctrl
2   5.58  ctrl
3   5.18  ctrl
4   6.11  ctrl
5   4.50  ctrl
6   4.61  ctrl
help(PlantGrowth)
Results from an Experiment on Plant Growth
Description
Results from an experiment to compare yields (as
measured by dried weight of plants) obtained under
a control and two different treatment conditions.
Usage
PlantGrowth
Format
A data frame of 30 cases on 2 variables.
[, 1] weight      numeric
[, 2] group factor
The levels of group are 'ctrl', 'trt1', and
'trt2'.

boxplot(weight ~ group, data=PlantGrowth,
col="blue")
```

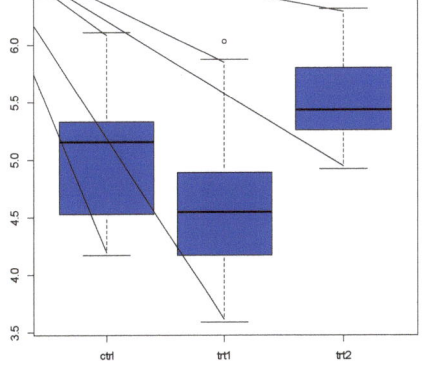

```
install.packages("lattice",repos="http://cran.r-
project.org")
library(lattice)
histogram(~ weight | group, data=PlantGrowth,
col="blue", type="count")
```

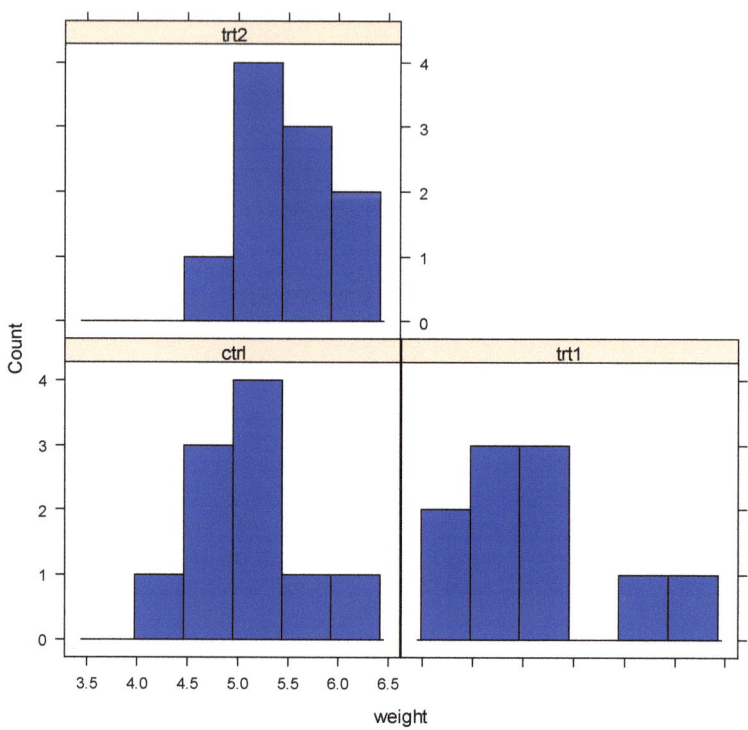

```
> summary(PlantGrowth)
     weight          group
 Min.   :3.590    ctrl:10
 1st Qu.:4.550    trt1:10
 Median :5.155    trt2:10
 Mean   :5.073
 3rd Qu.:5.530
 Max.   :6.310
```

```
> tapply(PlantGrowth$weight, PlantGrowth$group,
summary)
$ctrl
   Min. 1st Qu.  Median    Mean 3rd Qu.    Max.
  4.170   4.550   5.155   5.032   5.292   6.110

$trt1
   Min. 1st Qu.  Median    Mean 3rd Qu.    Max.
  3.590   4.208   4.550   4.661   4.870   6.030

$trt2
   Min. 1st Qu.  Median    Mean 3rd Qu.    Max.
  4.920   5.268   5.435   5.526   5.735   6.310

>  tapply(PlantGrowth$weight, PlantGrowth$group,
sd) #standard deviation
      ctrl        trt1        trt2
0.5830914 0.7936757 0.4425733

> weights <- aov(weight ~ group, data=PlantGrowth)
> summary(weights)
            Df Sum Sq Mean Sq F value Pr(>F)
group        2  3.766  1.8832   4.846 0.0159 *
Residuals   27 10.492  0.3886
---
Signif. codes:  0 '***' 0.001 '**' 0.01 '*' 0.05
'.' 0.1 ' ' 1
```

Interpretation: PlantGrowth data has 30 dry weights of plants that come from three groups: control, treatment 1, and treatment 2. From the box plots, we note that the largest weight in treatment1 has been shown as an outlier in the box plot. The histograms show that neither of the treatments follow a normal distribution. The summary statistics for each group were calculated. Visually comparing the two treatments to the control group might make us think that treatment 1 decreased the plant weight whereas treatment 2 increased it.

We used aov() to test this hypothesis H$_0$: all groups are equal, which conducts an F-test. More on test types on the next chapters. The small p-value makes us reject this hypothesis. The different standard variation for each group shows non-constant spread across groups. Because of the small sample size and difference in variance, we can argue that the result of the test could change if we have a larger sample from each treatment.

Example 6: Three Ordered Groups

```
data(warpbreaks)
> head(warpbreaks)
  breaks wool tension
1     26    A       L
2     30    A       L
3     54    A       L
4     25    A       L
5     70    A       L
6     52    A       L
> str(warpbreaks) # structure
'data.frame':    54 obs. of  3 variables:
 $ breaks : num   26 30 54 25 70 52 51 26 67 18 ...
 $ wool   : Factor w/ 2 levels "A","B": 1 1 1 1 1
1 1 1 1 1 ...
 $ tension: Factor w/ 3 levels "L","M","H": 1 1 1
1 1 1 1 1 2 ...
```
help(warpbreaks)
```
The Number of Breaks in Yarn during Weaving
Description
This data set gives the number of warp breaks per
loom, where a loom corresponds to a fixed length
of yarn.
Usage
warpbreaks
Format
```

A data frame with 54 observations on 3 variables.
[,1] breaks   numeric The number of breaks
[,2] wool   factor The type of wool (A or B)
[,3] tension factor The level of tension (L, M, H)
There are measurements on 9 looms for each of the
six types of warp (AL, AM, AH, BL, BM, BH).

```
> summary(fm1 <- aov(breaks ~ wool + tension, data
= warpbreaks))
            Df Sum Sq Mean Sq F value  Pr(>F)
wool         1    451   450.7   3.339 0.07361 .
tension      2   2034  1017.1   7.537 0.00138 **
Residuals   50   6748   135.0
---
Signif. codes:  0 '***' 0.001 '**' 0.01 '*' 0.05
'.' 0.1 ' ' 1
> TukeyHSD(fm1, "tension", ordered = TRUE)
  Tukey multiple comparisons of means
    95% family-wise confidence level
    factor levels have been ordered
Fit: aov(formula = breaks ~ wool + tension, data =
warpbreaks)
$tension
          diff        lwr       upr       p adj
M-H  4.722222  -4.6311985 14.07564 0.4474210
L-H 14.722222   5.3688015 24.07564 0.0011218
L-M 10.000000   0.6465793 19.35342 0.0336262

> plot(TukeyHSD(fm1, "tension"))
```

95% family-wise confidence level

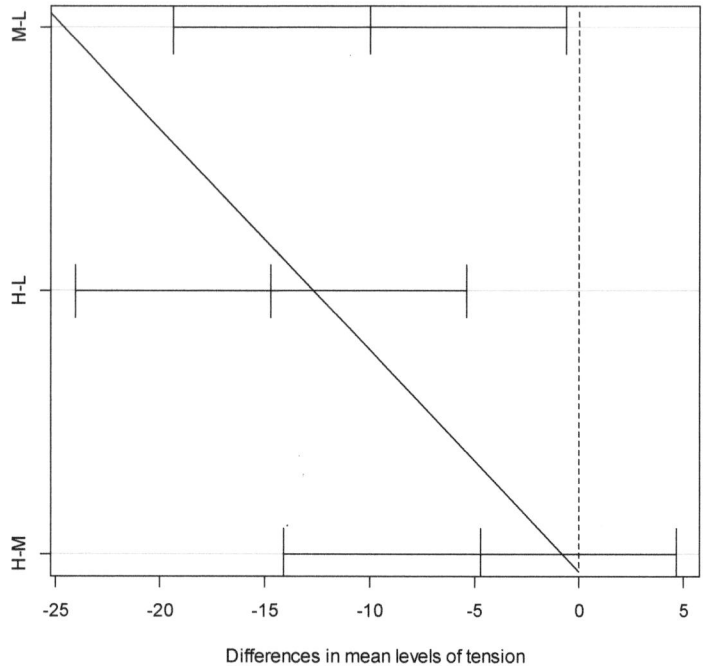

Differences in mean levels of tension

Interpretation: The tension levels (light, medium, high) of the wool are ordered meaning increasing in tension going from L to M to H. TukeyHSD() output shows that the differences L-H and L-M are significant while the difference M-H is not. The plot shows the same result where the confidence interval contains the zero meaning not statistically significant.

Chapter 3: Univariate and Multivariate Hypothesis Testing

All the hypothesis tests in the previous examples assume that the data follows a normal (Gaussian) distribution. This is an important assumption to linear and logistic regression models. If the data is not normal then a transformation is required. More on this on my book titled Statistics by Example.

Example 7: Univariate Normality Test

```
install.packages("datasets",repos="http://cran.r-
project.org")
library(datasets)
data(beavers)
> head(beaver1)
   day time   temp activ
1 346   840 36.33     0
2 346   850 36.34     0
3 346   900 36.35     0
4 346   910 36.42     0
5 346   920 36.55     0
6 346   930 36.69     0
help(beavers)
Body Temperature Series of Two Beavers
Description
```

Reynolds (1994) describes a small part of a study of the long-term temperature dynamics of beaver Castor canadensis in north-central Wisconsin. Body temperature was measured by telemetry every 10 minutes for four females, but data from a one period of less than a day for each of two animals is used there.
Usage
beaver1
beaver2
Format
The beaver1 data frame has 114 rows and 4 columns on body temperature measurements at 10 minute intervals.
The beaver2 data frame has 100 rows and 4 columns on body temperature measurements at 10 minute intervals.
The variables are as follows:
day
Day of observation (in days since the beginning of 1990), December 12-13 (beaver1) and November 3-4 (beaver2).
time
Time of observation, in the form 0330 for 3:30am
temp
Measured body temperature in degrees Celsius.
activ
Indicator of activity outside the retreat.
Note
The observation at 22:20 is missing in beaver1.

shapiro.test(beaver2$temp) # Measured body temperature in degrees Celsius
        Shapiro-Wilk normality test

data: beaver2$temp
W = 0.9334, p-value = 7.764e-05

<u>*Interpretation*</u>: For this test the null hypothesis H_0 is that the data comes from a normal distribution. Because p-value is lower than 0.05, we reject H_0.

Example 8: Multivariate Normality Test

This example tests the normality of multivariate data. In other words, it tests the null hypothesis H_0 of multivariate normality.

```
install.packages("mvnormtest",repos="http://cran.r
-project.org")
library(mvnormtest)
> data(EuStockMarkets)
> head(EuStockMarkets)
        DAX     SMI    CAC    FTSE
[1,] 1628.75 1678.1 1772.8 2443.6
[2,] 1613.63 1688.5 1750.5 2460.2
[3,] 1606.51 1678.6 1718.0 2448.2
[4,] 1621.04 1684.1 1708.1 2470.4
[5,] 1618.16 1686.6 1723.1 2484.7
[6,] 1610.61 1671.6 1714.3 2466.8
> help(EuStockMarkets)
Daily Closing Prices of Major European Stock
Indices, 1991-1998
Description
Contains the daily closing prices of major
European stock indices: Germany DAX (Ibis),
Switzerland SMI, France CAC, and UK FTSE. The data
are sampled in business time, i.e., weekends and
holidays are omitted.
Usage
EuStockMarkets
Format
A multivariate time series with 1860 observations
on 4 variables. The object is of class "mts".

> dim(EuStockMarkets)
[1] 1860    4
> mshapiro.test(t(EuStockMarkets)) # t() is the
transpose

        Shapiro-Wilk normality test

data:   Z
W = 0.9122, p-value < 2.2e-16
```

Interpretation: The low p-value tells us to reject H_0.

Example 9: Henze-Zirklers Multivariate Normality Test

```
install.packages("MVN",repos="http://cran.r-
project.org")
library(MVN)
hzTest(EuStockMarkets)
  Henze-Zirkler's Multivariate Normality Test
-----------------------------------------------
  data : EuStockMarkets

  HZ      : 37.71156
  p-value : 0

  Result  : Data are not multivariate normal.
-----------------------------------------------

uniPlot(EuStockMarkets, type = "qqplot")
```

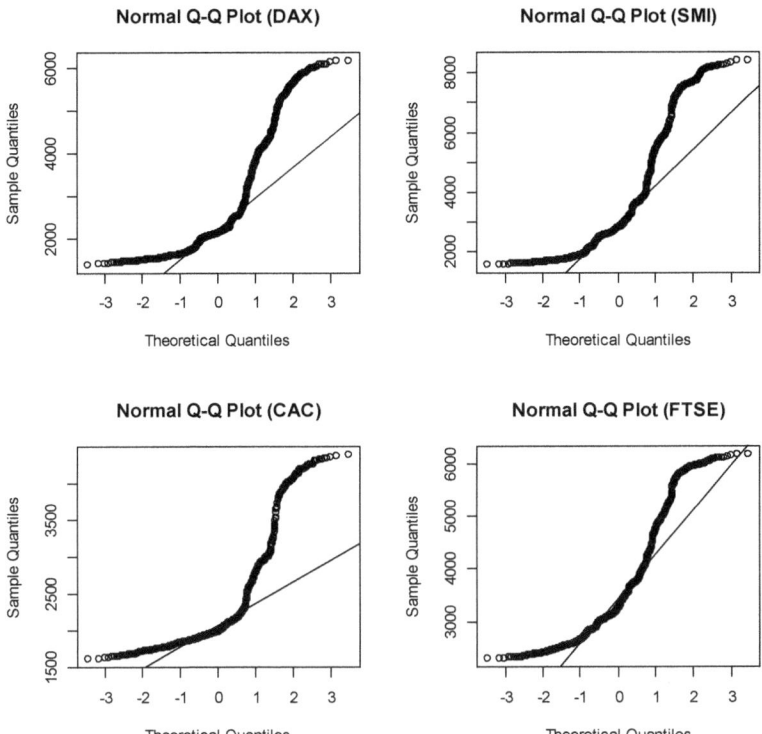

```
uniPlot(EuStockMarkets, type = "histogram")
```

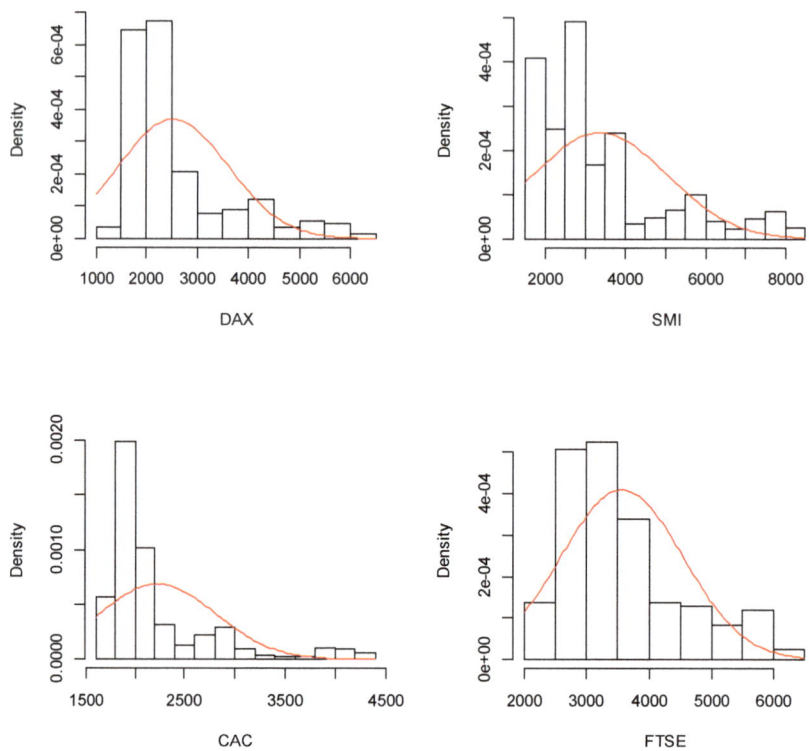

Interpretation: The low p-value tells us to reject H_0. The q-q plots and histograms demonstrate this conclusion as well. The data is far from multivariate normality. There are many other multivariate normality tests in R. More on univariate and multivariate hypothesis testing on the next chapter.

Chapter 4: Parametric and Non-parametric Hypothesis Testing

Based on the distribution assumption, there are two categories of hypothesis tests: parametric and non-parametric tests. Parametric tests assume the data follows a certain distribution, usually normal/Gaussian. Non-parametric tests do not assume that the sample is drawn from a specific distribution. The table below shows the most commonly used parametric tests and their alternative non-parametric ones. ANOVA stands for analysis of variance.

Parametric test	Non-parametric test
1-sample Z-test, 1-sample t-test	1-sample sign test
1-sample Z-test, 1-sample t-test	1-sample Wilcoxon
2-sample t-test (unpaired t-test)	Mann-Whitney test
One-way ANOVA (1 sample to k sample ANOVA)	Kruskal-Wallis test
One-way ANOVA	Mood's Median test
Two-way ANOVA	Friedman test
Pearson Correlation Coefficient	Spearman ~
Paired t-test	Wilcoxon Rank Sum

Since we cannot test for a categorical variable's distribution, we use non-parametric hypothesis testing. And parametric tests are usually used when having continuous data.

Example 10: ANOVA Test

```
>  data(InsectSprays)
> head(InsectSprays)
  count spray
1    10    A
2     7    A
3    20    A
4    14    A
5    14    A
6    12    A
> dim(InsectSprays)
[1] 72   2

>  str(InsectSprays) # structure
'data.frame':   72 obs. of  2 variables:
 $ count: num   10 7 20 14 14 12 10 23 17 20 ...
 $ spray: Factor w/ 6 levels "A","B","C","D",..: 1
1 1 1 1 1 1 1 1 1 ...

help(InsectSprays)
Effectiveness of Insect Sprays
Description
The counts of insects in agricultural experimental
units treated with different insecticides.
Usage
InsectSprays
Format
A data frame with 72 observations on 2 variables.
```

[,1] count numeric Insect count

[,2] spray factor The type of spray

```
> tapply(InsectSprays$count, InsectSprays$spray,
mean)
        A         B         C         D         E
F
14.500000 15.333333  2.083333  4.916667  3.500000
16.666667

# now let's check the variances
> tapply(InsectSprays$count, InsectSprays$spray,
var)
        A         B         C         D         E
F
22.272727 18.242424  3.901515  6.265152  3.000000
38.606061

#sample size
> tapply(InsectSprays$count, InsectSprays$spray,
length)
 A  B  C  D  E  F
12 12 12 12 12 12
boxplot(count ~ spray, data=InsectSprays)
```

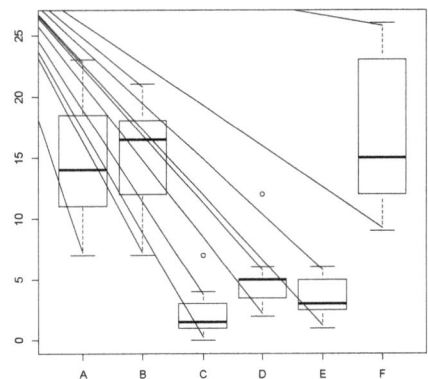

```
>  oneway.test(count ~ spray, data=InsectSprays)
        One-way analysis of means (not assuming
equal variances)

data:  count and spray
F = 36.0654, num df = 5.000, denom df = 30.043, p-
value = 7.999e-12
```

Interpretation: The test evaluates which spray is better by statistically comparing the numberof insects for each type. oneway.test assumes the samples are from normal distributions. The variances are not necessarily assumed to be equal. Because of the small p-value we reject the null hypothesis H_0 that the sprays have similar effects.

Example 11: Kruskal-Wallis Non-parametric Test

```
help(airquality)
```
New York Air Quality Measurements
Description
Daily air quality measurements in New York, May to
September 1973.
Usage
airquality
Format
A data frame with 154 observations on 6 variables.

[,1]	Ozone	numeric	Ozone (ppb)
[,2]	Solar.R	numeric	Solar R (lang)
[,3]	Wind	numeric	Wind (mph)
[,4]	Temp	numeric	Temperature (degrees F)
[,5]	Month	numeric	Month (1--12)
[,6]	Day	numeric	Day of month (1--31)

```
> head(airquality)
  Ozone Solar.R Wind Temp Month Day    newO3
1    41    190   7.4   67     5   1 6.403124
2    36    118   8.0   72     5   2 6.000000
3    12    149  12.6   74     5   3 3.464102
4    18    313  11.5   62     5   4 4.242641
5    NA     NA  14.3   56     5   5       NA
6    28     NA  14.9   66     5   6 5.291503
boxplot(Ozone ~ Month, data = airquality)
```

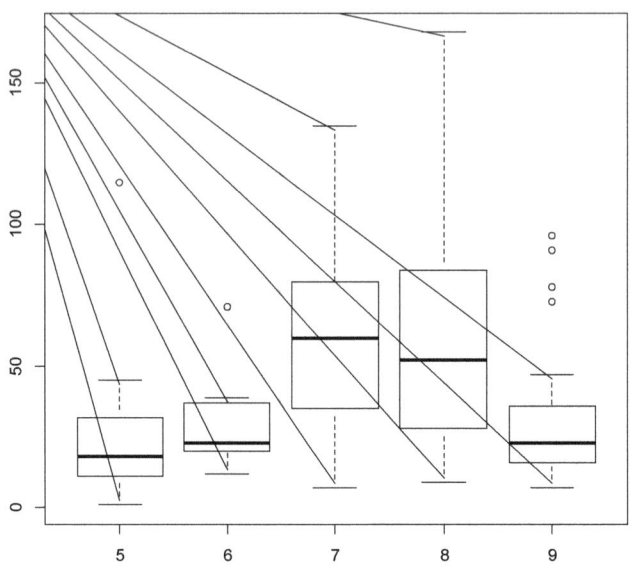

```
> kruskal.test(Ozone ~ Month, data = airquality)
         Kruskal-Wallis rank sum test

data:  Ozone by Month
Kruskal-Wallis chi-squared = 29.2666, df = 4, p-
value = 6.901e-06
```

Interpretation: The data has ozone levels sampled in NY in five consecutive months. The low p-value tells us to reject H_0 that ambient ozone levels were the same from May to September, 1973.

Example 12: MANOVA Test

```
library(datasets)
data(iris)

help(iris)
Edgar Anderson's Iris Data
Description
This famous (Fisher's or Anderson's) iris data set
gives the measurements in centimeters of the
variables sepal length and width and petal length
and width, respectively, for 50 flowers from each
of 3 species of iris. The species are Iris setosa,
versicolor, and virginica.
Usage
iris
iris3
Format
iris is a data frame with 150 cases (rows) and 5
variables (columns) named Sepal.Length,
Sepal.Width, Petal.Length, Petal.Width, and
Species.
iris3 gives the same data arranged as a 3-
dimensional array of size 50 by 4 by 3, as
represented by S-PLUS. The first dimension gives
the case number within the species subsample, the
second the measurements with names Sepal L., Sepal
W., Petal L., and Petal W., and the third the
species.
Source
Fisher, R. A. (1936) The use of multiple
measurements in taxonomic problems. Annals of
Eugenics, 7, Part II, 179-188.
The data were collected by Anderson, Edgar (1935).
The irises of the Gaspe Peninsula, Bulletin of the
American Iris Society, 59, 2-5.
```

```
> head(iris)
  Sepal.Length Sepal.Width Petal.Length Petal.Width
Species
1        5.1          3.5          1.4          0.2
setosa
2        4.9          3.0          1.4          0.2
setosa
3        4.7          3.2          1.3          0.2
setosa
4        4.6          3.1          1.5          0.2
setosa
5        5.0          3.6          1.4          0.2
setosa
6        5.4          3.9          1.7          0.4
setosa

> names(iris) <- c("SL", "SW", "PL", "PW", "SPP")

> # fit a multivariate one-way ANOVA model to the
iris data
> mod.iris <- lm(cbind(s) ~ SPP, data=iris)
> mod.iris

Call:
lm(formula = cbind(SL, SW, PL, PW) ~ SPP, data =
iris)

Coefficients:
                SL       SW       PL       PW
(Intercept)   5.006    3.428    1.462    0.246
SPPversicolor 0.930   -0.658    2.798    1.080
SPPvirginica  1.582   -0.454    4.090    1.780

> anova(mod.iris)
Analysis of Variance Table

             Df  Pillai approx F num Df den Df    Pr(>F)
(Intercept)   1 0.99313   5203.9      4    144 < 2.2e-16
***
SPP           2 1.19190     53.5      8    290 < 2.2e-16
***
Residuals   147
---
Signif. codes:  0 '***' 0.001 '**' 0.01 '*' 0.05 '.' 0.1
' ' 1
```

Interpretation: The data has measurements (in cm) of the variables sepal length and width and petal length and width, respectively, for 50 flowers from each of 3 species of iris. The multiple (multivariate outcome) regression model mod.iris estimates the effect of SPP on the four variables SL, SW, PL and PW simultaneously. Then the function anova() tests the hypothesis (H_0) that the four response' means are identical across the three species of irises. The small p-value advise against H_0.

 The End